# Cedar Key Fiber And Brush Factory

*"From The Tree To Thee"*

By John W. Andrews, M.D.

**Cedar Key Fiber and Brush Factory**



Published by Atlantic Publishing Group, Inc.
1405 SW 6th Avenue • Ocala, Florida 34471 • Phone 800-814-1132 • Fax 352-622-1875
Website: www.atlantic-pub.com • Email: sales@atlantic-pub.com
SAN Number: 268-1250



Library of Congress Cataloging-in-Publication Data

Andrews, John W., 1929- author.
Cedar Key Fiber and Brush Factory : from the tree to thee / John W. Andrews, M.D.
pages cm
ISBN 978-1-62023-130-2 (alk. paper) -- ISBN 1-62023-130-1 (alk. paper) 1. Cedar Key Fiber and Brush Factory. 2. Broom and brush industry--Florida--Cedar Key--History--20th century. 3. Fiber plants--Florida--Cedar Key--History--20th century. I. Title.
HD9971.5.B764C4325 2015
338.7'6335--dc23
2015033301



Cover & Interior Design: Meg Buchner • megadesn@mchsi.com
Cover Painting Courtesy of Jan Parkinson

Printed in the United States

*Finally, a legitimate history of The Standard Manufacturing Company of Cedar Key, Florida!*

*Learning about the company and the people behind this business is a fascinating read. You will become familiar with terms such as cabbage palms, boots of the palm, hackles, barges that collected the cut palms, making of brooms, where around this world the fiber or brooms were shipped, how the whole idea of making brooms out of the fiber of the cabbage palm came about and of course the founder of the Company, Dr. Dan. I wish I could talk with Dr. Dan and ask him questions. What an interesting man. Trained as a dentist and opens a factory that makes brooms out of the fiber of a palm tree and hires 100 people that live on the island. A huge boost to the economy of the small coastal town and a company that stayed productive during the Great Depression. Amazing story.*

*This book is a good read telling the interesting and accurate history of a vital business that produced a quality product in the small Gulf Coast town of Cedar Key, Florida. Meet Dr. Dan, Cora, Dan Jr., and Dr. John and learn how The Standard Manufacturing Company came to be, what it produced, what it meant to the town of Cedar Key and how Dr. John still makes brooms using the same method as the factory did and with the fiber that was left in a warehouse after the closing of the factory. Enjoy!*

— Minnie Crevasse

*This is an extremely authentic account of the production of fiber and brushes from the cabbage palm tree.*

—Sally Baylor

---

*Just finished reading the history of the Fiber Factory in Cedar Key, Florida. Very Interesting! The author did not have to do research to write the book – he had "**Lived It**".*

*It has never ceased to amaze me how the human mind works. How could anyone look at the boot of a cabbage palm and visualize the many processes it would have to go through to finally become a beautiful brush?*

*I am sure Dr. Dan and Dan Jr. – because of their love for the "Old Factory" – would be pleased to know their part of Florida history will be preserved in book form for future generations. It is a thing of the past and will never be repeated. Good job, Johnny!*

—Wilma (Mrs. Dan Jr.) Andrews

*This is absolutely the best book I have read about Cedar Key since it includes historical information of this period and the everyday life of people in this community from the earliest part of the 1900's up until the 1950's.*

*I met Dr. John Andrews in 2001 when we moved here for me to work as a minister. George Delaino and his wife Laura Jean quickly became our close friends, and they introduced my wife and I to John Andrews. I always listen with wide-eyed attention each time John tells visitors at the Cedar Key Historical Society Museum about his boyhood memories of growing up on this island.*

*If you are interested in Cedar Key, you will likely be the same as me – I could not put the book down and had to go back and be sure of some very interesting detail – like the war tax on wages in WWII.*

— David Binkley, Sr., Minister

# DEDICATION

This book is dedicated to my mother, Cora Lee Tooke Andrews (deceased). Her love for her family and the Fiber and Brush factory is reflected in what she saved after the factory closed and my father died. She lived to the ripe old age of 98, passing away in 1992. She is buried in the family plot in Cedar Key Cemetery.

She saved a wealth of family and factory photos, correspondence, ledgers, and other items from the office as well as many artifacts from the factory itself. She even saved some fiber left in an old warehouse after the factory closed and it is on display in the Cedar Key Historical Museum.

Without her thoughtfulness it would not have been possible to write this book nor create the Fiber and Brush Factory exhibit in the Andrews House at the Cedar Key Historical Museum.

# TABLE OF CONTENTS

# ACKNOWLEDGEMENTS

I have thought for a long time about writing a book about my family and the little known fiber and brush factory in Cedar Key. My friends have prodded me to do it but it took my wife Joanne to finally get me to do it. Also without her tremendous help it never would have happened. She is a very organized person who recorded the text, helped me organize it and put all of the pictures together in an orderly fashion. She devoted many hours dealing with my changes to the text and what pictures to use and where to put them. Without Joanne, it never would have happened. I am thankful to her and give her most of the credit.

Thanks a bunch to my cousin, Sally Baylor. Her mother, my Aunt Virginia, was the secretary and bookkeeper for the Company for many years. Sally is a retired schoolteacher who edited and re-edited the text and corrected many grammatical and typographical errors. She used a red pencil freely and also made several helpful suggestions.

A huge thank you goes to Elizabeth Ehrbar and helpers Peggy Rix for putting together the Fiber and Brush Factory exhibit in the Andrews House. Thanks also go to Maurice Rix and Gil Ruggles for building some of the display cases.

Thanks go to artist Bill Roberts for the beautiful Fiber Factory Exhibit mural depicting the harvesting and transportation of the cabbage buds and for allowing me to use photos of this in the book. Thanks also to Wayland Wadley for creating a detailed replica of the factory site of which a photo of this is also included in the book.

I wish to also thank my sister-in-law, Wilma Andrews, Sally Baylor, David Binkley, and Minnie Crevasse for reviewing the book, and thank you to Dr. Bob McCollough for his generous foreword.

Joanne's daughter, Dini Vascellaro's technical support was invaluable and greatly appreciated.

Thank you to the team at Atlantic Publishing: Doug Brown for his great interest and effort in making this book possible; Crystal Edwards for helping make the publication process

so simple. A special thanks to Meg Buchner for the cover and her artistic designs.

I also thank my daughters Jennie, Debbie, Vivian, and granddaughter Kelley, who have hoped for this book for a long time and finally it is one of my legacies to them. Jennie actually came up with "From "The Tree to Thee" as part of the title.

I cannot fail to thank all of the hardworking workers that performed the many difficult jobs involved in the factory operation. Most of the jobs were very hard, the hardest being the cutting and hauling of the buds in the swamp especially dealing with the heat and exposure to the insects. I doubt that anyone would be willing to do that kind of work today.

# FOREWORD

Americans today are generally aware of the large corporations that dominate our industry. But in spite this age of search engines we remain largely unaware of the myriad small industries that underpin our economy. We are even less aware of historical industries that gave employment to millions of workers only to succumb to advanced technologies in the last century.

The cabbage palm fiber industry, so accurately described herein by Dr. Andrews serves as a reminder of the endeavors of one of the least known industries of the past century.

It is the story of how a tiny coastal town was sustained by ingenuity and difficult labor, to survive in spite of the Great Depression, World War II, hurricanes and biting insects.

I had the privilege of practicing medicine with "Dr. Johnnie" (as he is affectionately known in Cedar Key) for 25 years and he continues to amaze me with endeavors like this narrative which promote the history of his birthplace.

– Robert H. McCollough, M.D.

# INTRODUCTION

This book is about a little known old Florida industry and about the family who created it. Located in the small Florida Gulf Coast town of Cedar Key. It was known locally as the "Fiber Factory". The factory operated from 1910 to 1952 except for several years during and after World War II when it was closed due to a shortage of workers. It was one of the two major industries in Cedar Key. The factory and the seafood industry helped sustain the town economically during the above years. The palm fiber industry was unusual because Florida was the only state in which it existed. There were only three factories: one in Jacksonville owned by the Wooten family, and another near Sanford owned by the Ox Fiber Brush Co., of Baltimore, Maryland. The Cedar Key Factory closed in 1952, and the others closed several years later because they could not compete with the cheap synthetic fibers that had been introduced. The palmetto fiber, as it was called, was much more expensive to produce because it required such a labor-intensive process.

The fiber came from the Sabal Palm commonly known as the cabbage palm that is now the Florida State Tree. The Donax brushes made from it are now relics of the past.

I have included a great deal about my family since so many of them were a part of it. I have also included the names of many who worked in the factory over the years because some of their descendants may be interested in knowing.

There is a comprehensive exhibit of the Cedar Key Fiber and Brush Factory in the Cedar Key Historical Museum. It is located in the Andrews Home. The Andrews home was donated to the Cedar Key Historical Society Museum in 1995. It was moved from its original site near the Factory (where the Fenimore Mill Condominiums are now located) to the Museum property and is now part of the Museum. This exhibit was made possible because of the many items and photos that my mother saved and stored in her garage. The Museum is also much indebted to Mrs. Elizabeth Ehrbar, our volunteer curator and her helper Peggy Rix who put the exhibits together. Anyone reading this book can gain a better understanding of the things described herein by visiting the museum. The house and its exhibits were opened to the public in 2001.

We are also indebted to the volunteer workers that helped with the restoration of the building. These included my brother Dan Jr., Kenny Baylor, Burton Walrath, and especially George Delaino. George was an excellent finished carpenter who replaced the deteriorated interior woodwork and sanded all of the window and door facings and then reinstalled them. The window and door facings had been put up with brass screws making it possible to remove them, sand them and reinstall them. George even sanded the heads of the screws and used them to re-install the woodwork. With the help of the other volunteers George did all of this. Greg Lang was the volunteer licensed general contractor who did a great job and to whom we are greatly indebted as well.

When the house was moved from its original location on the Fenimore Mill site to the museum property, Central Florida Electric, Southern Bell Telephone Company and Brighthouse Cable Company all donated their services which were required to raise power lines, telephone lines and cables to allow the two story house to pass under them. The major portion of the funding to restore the building came from a grant from the State of Florida Division of Cultural Affairs. Additional funds were needed and were provided in the form of donations. Donations were received from many local Cedar Key residents and businesses as well as some major donations from friends of mine in Gainesville - Dr. Lamar Crevasse, Marvin Gresham, and Al Alsobrook. Yard sales and fish fry fundraisers were also held. Don Lindsey of Chiefland prepared the foundation at a greatly reduced cost to us.

None of this would have been possible without the help of the State Department of Historic Preservation and Cultural Affairs in Tallahassee and all of the many donations by individuals, businesses and companies who wanted to help preserve a part of Cedar Key's unique history.

# FAMILY HISTORY

The first history we have of the Andrews family, presumably from Wales, is that of Christopher Andrews living in White Hall Town, NY in the 1790 census. He had ten children. His son Samuel married Sylvia Harris, daughter of Joshua Harris, a Revolutionary War veteran. She was a granddaughter of Moses Harris who served as a spy under George Washington who was given a citation by General Washington and a grant of 1,000 acres of land in New York State. Samuel and Sylvia Andrews had seven children. One son, Daniel H. Andrews, became a physician. He came to the Florida territory following his medical education in Cincinnati in the 1830's during the second Seminole Indian War. He practiced medicine on horseback in the northern part of Florida, then returned to Muncie, Indiana in 1842. He married Mary Jane Gilbert, daughter of Goldsmith Gilbert in 1844 in a ceremony by a Reverend Irwin, the first Presbyterian minister in Muncie. Mary Jane Gilbert was the first white child born in Delaware County, Indiana. Daniel and Mary Jane Andrews had four children. My grandfather, John Edwin, married Hanna Yingling in 1874. They had four sons, George Rex, Daniel Alfred, Sr., C. Raymond, and Forrest. He ran a flour mill in Muncie. After his wife died in 1908 he moved to Cedar Key and lived with my parents until he died at the age of 81 in 1930. I never knew him because I was only 1 year old when he died. My Father died at age 83 and my mother lived to be 98. They are buried in the Tooke family lot in Cedar Key. My brother Dan Jr. who passed away in 2003 is buried in Ocala where he lived most of his life.

*Andrews Brothers and their Father*
*Sitting (left to right): Dr. George and their father John E. Andrews*
*Standing: Dr. Daniel A. Sr., Raymond, and Forrest Andrews*

Dr. Daniel Alfred, Sr. was 31 when he began operating the factory. He remained single until he married my mother, Cora Lee Tooke, in Cedar Key in 1917. He was 39 years old and my mother, Cora Lee Tooke Andrews was 23. They lost their first child Helen Marjorie at age four to pneumonia in 1922. This, of course, was devastating especially to my mother. My brother Dan, Jr. was born in 1924 and I was born in 1929 when my dad was 52.

My mother's family goes back to the 1850's in Cedar Key. They were from England. Jesse Britt Tooke came to Virginia in 1622 and was a member of the House of Burgesses from Isle of Wight County, Virginia in the 1630's. Our branch of the Tooke family migrated south through the Carolinas to Cedar Key via Madison, Florida. They were all farmers until they arrived in Cedar Key when they became commercial fishermen. My great grandfather John Gideon Tooke married Choley Ellen Wilson (daughter of the first lighthouse keeper on Seahorse Key) in 1854. He was a commercial fisherman and a hunting guide. He was a private in the Florida Infantry CSA during the Civil War and received a 160-acre land grant in Levy County in 1862.

My grandfather, John William Tooke, a commercial fisherman married Sara Jane Delaino in 1889. They both had rather short lives dying at the ages of 57 and 46 respectively. They had 15 children, three of whom died in infancy and one died accidentally at age 15. The youngest ones, being left without parents, were raised by my three oldest aunts. Four of my aunts and two of my uncles lived well into their nineties. Most of them are buried in the Cedar Key Cemetery. My mother and three aunts were born on Scale Key. An aunt and an uncle were born on Atsena Otie. All of the others were born on Way Key (Cedar Key).

*Left to right:*
*Sarah Jane Delaino Tooke, Grandmother;*
*John William Tooke, Grandfather;*
*Delilah Hattie Mae Tooke, Aunt*

My mother's old family home "Tooke House" was located next door to the Andrews House which was located next to the Fiber Factory. The local school had one hour for lunch and my brother Dan, myself and six cousins alternated eating lunch (we called it dinner) either at the Tooke House prepared by my aunts or at our house prepared by my mother. It was only a mile to the school house so we had time to eat and walk or ride bicycles and get back to school on time.

*Naomi Tooke Dorsett*

*Tooke Family Aunts and Uncles (left to right):*
*Sitting: John Tooke, William Tooke, Virginia Tooke Turner Pugh, Jesse Tooke*
*Standing: Agnes Tooke, Cora Tooke Andrews, Esther Tooke, Ellen Tooke*

*Andrews Family (left to right):*
*Dr. Dan, John, Cora, Dan Jr.*

*Uncle George Furney Tooke*
*Commercial Fisherman*
*and Boat Builder*

# FOUNDING THE COMPANY

Dr. Daniel A. Andrews, Sr., was born and raised in Muncie, Indiana. He attended dental school at the University of Indianapolis Dental College and graduated in 1899. He developed a flourishing practice in Indianapolis where he practiced primarily restorative dentistry including inlays, crowns, and even root canals as shown in one of his ledgers in 1904. Charges recorded were 50 cents for amalgam fillings, $4.50 for gold foil fillings, and $4.50 for root canals.

He came to Cedar Key on the train on a hunting and fishing vacation in 1904. While here he met Mr. St. Clair Whitman. There may have been another factory in Cedar Key that Mr. Whitman had worked in. There is an old postcard picture labeled Cedar Key Fiber Factory, but I have been unable to find any other reference to a fiber factory prior to the one Dr. Dan built. He was called Dr. Dan by everyone in Cedar Key. At any rate Mr. Whitman seemed to have knowledge of a "hackle" machine that could be used to extract fiber from the cabbage palm.

Mr. Whitman was a man of many talents. Besides being a mechanical genius he was a well-known collector of seashells, Native American Artifacts and butterflies. His seashell collection came from different parts of the world in addition to the ones he collected in Cedar Key and around Florida. Some of these are on display in his old home at the Cedar Key State Museum.

*St. Clair Whitman beside his 1925 Chevrolet parked in front of the Tooke House. He drove it until his death in 1959 at the age of 91.*

Dr. Dan became so interested, that after several subsequent trips to Cedar Key he decided to give up his dental practice in Indianapolis and move to Cedar Key. With the help of

Mr. Whitman he built a fiber and brush factory. I, and others, have often wondered why he made such a drastic change in his life's work. I can only guess, but several factors may account for it. He grew up around his father's steam-operated flourmill in Muncie, Indiana and the fiber factory he built in Cedar Key was also a steam-operated mill. He contracted pneumonia on two occasions during Indiana winters and liked the thought of living in a much warmer climate, especially in Cedar Key that he had come to love. He lost his first wife in Indiana from acute leukemia after only four months of marriage. Last, he may have seen himself becoming bored, doing nothing but dentistry the rest of his life. He obviously was adventurous by nature.

*Founders of the Company*
*Top left: Dr. Dan*
*Top right: Ray Andrews*
*Bottom left: Dr. George Andrews*
*Bottom right: Forrest Andrews*

After a number of meetings with Mr. Whitman in Cedar Key he proceeded to form the Standard Manufacturing Company. There were three stockholders: Dr. Dan, Mr. John

Chapman and Dr. Homer Allen, all from Indianapolis. Each held 1/3 of the stock with a capitalization of $10,000. Later Dr. Dan's three brothers became stockholders with the issuing of an additional $5,000 of stock. Over the years Dr. Dan bought out all of the stockholders except for his brother, Forrest, who moved to Cedar Key and helped him operate the factory.

*Factory building shortly after completion in 1910. Dr. Dan and an unidentified person on the steps.*

Construction was completed and the Standard Manufacturing Company began operating in January 1910. It operated until 1952 except for a 6-year period from 1944 until January 1950 when there were not enough workers available. Most of them had entered the military service during World War II and many of them did not return to Cedar Key after the war.

My brother, Dan Jr. after serving in World War II finished college with a business degree from the University of Florida and returned to Cedar Key. He and our Dad refurbished the factory, hired enough workers and re-opened the factory in January 1950.

*Damage to the Fiber Factory Building after the 1950 hurricane.*
Courtesy: Andrews House Exhibit

Things were going well producing fiber and brushes again until a destructive category 3 hurricane hit Cedar Key in September 1950. It destroyed four of the seven buildings that made up the factory complex and they had no hurricane insurance. It continued operating on a limited basis producing fiber and brushes until 1952 when its doors were closed for good.

*Destruction of the "Broom Room" after the 1950 hurricane.*
In the Andrews House Exhibit

There were two other factories in Florida, one in Jacksonville and one near Sanford, Florida. Both of those plants closed a few years later because they could not compete with the

much cheaper synthetic fibers that were introduced. It is unlikely that this wonderful fiber will ever be produced again because it requires such a labor intensive and expensive process.

My brother Dan left Cedar Key and worked as superintendent in the Wet Department of the factory near Sanford for a couple of years. He then moved to Ocala and joined an accounting firm there where he spent the rest of his life. He passed away in 2003 at age 79 and is buried there.

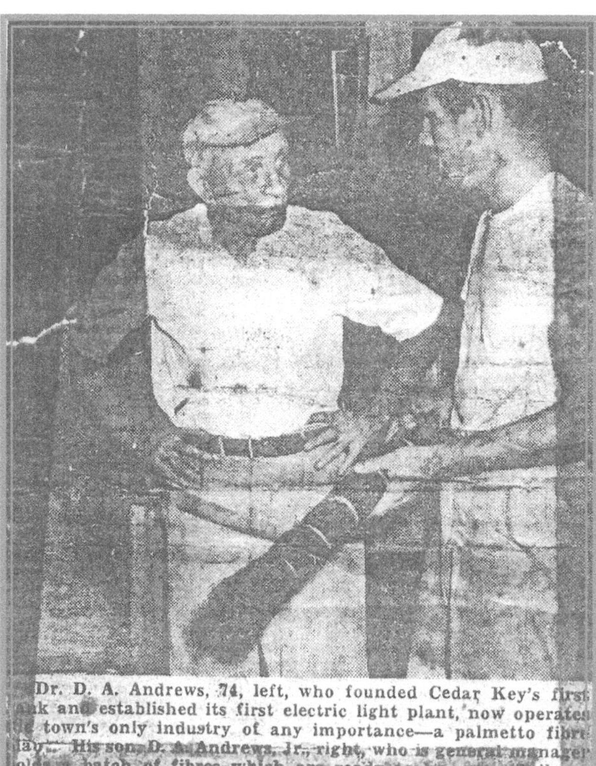

Dr. D. A. Andrews, 74, left, who founded Cedar Key's first [illegible]nk and established its first electric light plant, now operates [illegible] town's only industry of any importance—a palmetto fibre [illegible]ay. His son, D. A. Andrews, Jr., right, who is general manager [illegible]olds a batch of fibres which are ready to be cut.—(Tribune [illegible])

*Newspaper clipping showing Dr. Dan and Dan Jr. standing in the Factory discussing things. Dan Jr. is holding a bundle of "natural length fiber".*

*Dr. Dan and Dan Jr. standing in the drying yard with Andrews House in the background.*

*Dr. Dan*

# FACTS ABOUT THE SABAL (CABBAGE) PALM

The cabbage (Sabal) palm from which the fiber was produced was designated the Florida State Tree in 1953 and has replaced the coconut palm on the Florida State Seal. It is a very interesting tree that has had a number of uses over the years. The palm heart is edible and was used for food by the Seminole Indians. It was called "swamp cabbage" and eaten by early Florida settlers as well. It is still so-named and eaten by many Floridians who cook it today as you would regular cabbage. The heart is also served raw as "heart of palm salad" in some of the restaurants along the Florida Gulf Coast. It was first served by Bessie Gibbs at the Island Hotel in Cedar Key in the 1940s.

The fronds of the cabbage palm were used by the Seminoles for the roofs of their chickee huts. Some of the bud cutters also built temporary camps in the swamp and used the fronds for their thatched roofs.

In the early 1900's The Standard Manufacturing Company shipped shredded palm fronds in bales called "green fiber" to the Federal Penitentiary in Atlanta, Georgia where it was used for mattress filling. This may have come from the "Old Fiber Factory" on Cow Creek (a tributary of the Waccasassa River in Gulf Hammock). Little or nothing is known about that operation.

After the tree has grown up into a tall palm the trunks were used as pilings, especially for the construction of docks. This was a popular use in the 1800's and early 1900's. Many

*A typical thicket of cabbage palms.*

were sold and shipped by the Hodges Family from their land in southern Levy County to build the docks in Key West. The docks of the Standard Manufacturing Company as well as some other docks in Cedar Key also used cabbage pilings. Metal caps were put on the tops of the pilings to keep water out and prevent rotting. These would then last for 30 years or more. Unlike wooden pilings, they were resistant to marine worms that destroyed the underwater portions of the wooden pilings.

Even though the Fiber Factory used two to three thousand trees per month, there was never a shortage. It was not necessary to replant. The palms grew abundantly, almost like weeds, along the Gulf Coast. The fiber came only from the trunks of the small immature trees and seeds falling from the tall palms was a natural re-seeding process. Since only the small trees (selective cutting) were used, the same territory could be cut every 3 to 4 years and there was never a shortage. Each tree produced about 3 pounds of fiber.

These trees are just as thick now as they were then along the Big Bend's Gulf coast where the trees for the factory came from. The owners were paid stumpage (a certain amount for each tree) so it was a win-win operation.

The cabbage palm is no longer used to make fiber. The most common use today is for landscaping around homes, commercial buildings and along highways. Because the mature, seed-producing trees are used for this purpose, it is possible they could become less abundant in the future although they are so prolific it is hard for me to imagine that.

The berries of these trees also provide food for a variety of wildlife including raccoons, bears, and many species of birds. Bears can even tear small palms apart and eat the hearts.

*Cabbage (sabal) palm is called a "bud" after it is harvested.*

# THE FACTORY SITE

The factory site was located at the eastern tip of Cedar Key known as the Fenimore Mill Point located at the end of Second Street. It was originally the site of the Fenimore Steam Saw and Planing Company that was destroyed in the 1896 hurricane. The Fenimore Mill site was later owned by Mr. W. R Hodges the father of State Senator Randolph Hodges of Cedar Key. The Standard Manufacturing Company initially leased this 5 ½ acre site from Mr. Hodges with an option to buy. The factory was built in 1909 and began operating in 1910. The site was not actually purchased until 1917. The property is now occupied by the Fenimore Mill Condominiums.

Before the property was leased by the Standard Manufacturing Company, Mr. Hodges had acquired and moved by barge, about 15 abandoned houses which had survived the 1896 hurricane on Atsena Otie, the site of the first town of Cedar Key. The houses were set up on Fenimore Mill Point and rented to local residents. Mr. Hodges moved

*Standard Manufacturing Company's Cedar Key Fiber Factory. The building under construction in 1909.*

Courtesy: Bill Bale

them again to other properties he owned so the fiber factory could be built there. One small three-bedroom house that later became the first part of the Andrews House was retained by Dr. Dan Andrews. This building was the office for the Standard Manufacturing Company and also served as his residence. Another building called the "long house" was also retained. After a second story was added it became the building where the brushes were made and it was called the "broom room".

# THE FACTORY COMPLEX

1. **The Main Factory Building.** This two story metal building was about 125 feet long and 40 feet wide. It contained a central line shaft through its center the length of the building. All of the hackle machines both upstairs and downstairs ran off it by means of belts and pulleys. It was a steam-powered mill using wood for fuel.

2. **Oil Treatment Building.** This was a single story metal building about 100 feet long and 35 feet wide. It contained a large oil reservoir for oiling the fiber, and an attached inclined metal ramp for draining the fiber.

3. **Brush Factory (BROOM ROOM).** This two story wooden building was used for making the brushes. It contained long tables downstairs where women worked at an assembly line making brushes. The specially selected fiber for the brushes was stored upstairs in the building.

*The "Broom Room" Building where the brushes were made is pictured in the background. In the foreground is the metal shed where barges were built.*

4. **The Clock House.** This small building contained a time clock. Each worker had a card kept there. The worker would insert the card into a slot and the time would be recorded along with the day each time the worker "clocked" in and out.

*Left to right: Fiber Factory with workers standing in front. The small house in front of the factory building is the Clockhouse. Warehouse and railroad tracks are in the right foreground and the water tower is in back of the #2 warehouse.*

5. **Warehouses** included a single story metal building about 60 feet long and 30 feet wide; a single story concrete warehouse about the same size; and a two-story concrete block warehouse about 70 feet long and 35 feet long. These buildings were

used to store fiber and were called Warehouse numbers 1, 2, 3. After the factory closed in 1952, number 3 warehouse was leased to a Crab Canning Company from Maryland for a number of years. Now it houses the office for The Fenimore Mill Condominium Association and is the only building remaining. There was a 40 foot steel tower beside Warehouse 2 with a 5,000-gallon water tank on top. There was a well from which water was pumped to the tank and then gravity fed it to the factory.

6. **Dock for Boats and Boat Supply House.** A channel ran behind the factory on its way to Number 2 Bridge (Canal Bridge) on Highway 24. On the south edge of this channel, near the north end of the factory building was a wooden dock and boat supply house. The dock was about 50 feet long by 15 feet wide. The two towboats could be tied there and a row of cabbage pilings was provided for the barges to be tied.

Part of the dock was occupied by a boat supply house about 25′ by 15′ in size. Anchors, ropes anchor chains, life preservers, hand pumps, buckets, fuel, etc., were stored there.

The factory maintained a set of ways for boat and barge maintenance. This consisted of railroad rails going down into the water. There was a cradle onto which boats were driven and a gasoline-powered wench pulled them up out of the water on the cradle. Barnacles growing on the bottoms of the boats were a big problem despite use of anti-fouling copper paint. Boats and barges were pulled out and scraped every four to six months.

*Fiber Factory Building with a stack of buds ready to be boiled. The water tower and Andrews house is to the right.*

## FIRE HAZARD AT THE FIBER FACTORY

The floors and tables in all of the Fiber Factory buildings became saturated with the oil used for treating the fiber. These included the main factory building; three warehouses where fiber was stockpiled and the "Broom Room" where brushes were made. "No Smoking" signs were everywhere because of the risk of a disastrous fire. Anyone caught smoking on the premises was fired immediately. The factory had its own mobile chemical fire tank on wheels that was kept in a small building on site. There was no fire department in Cedar Key at the time so the factory had to provide its own fire protection.

An issue in this regard involved the power and ice plant (owned by Florida Power and Light Co) located next to the Fiber Factory property. On occasion the flue from the stack of the diesel engine would become partially clogged with hot residue. This would become dislodged and wind would deposit it on the factory property, sometimes in areas covered by waste fiber. This could easily start a fire because it was quite flammable when dry. This did not happen if the flues were cleaned at regular intervals. Florida Power was put on notice that if this caused such a fire they would be held legally liable. Fortunately, a serious fire never occurred.

*Fiber Factory Building which contained the Wet Department downstairs and the Finishing Department upstairs. This also shows the Drying Yard and the water tower.*

*This Andrews House Exhibit, created by Wayland Wadley, shows the buildings and layout of the Cedar key Fiber Factory.*

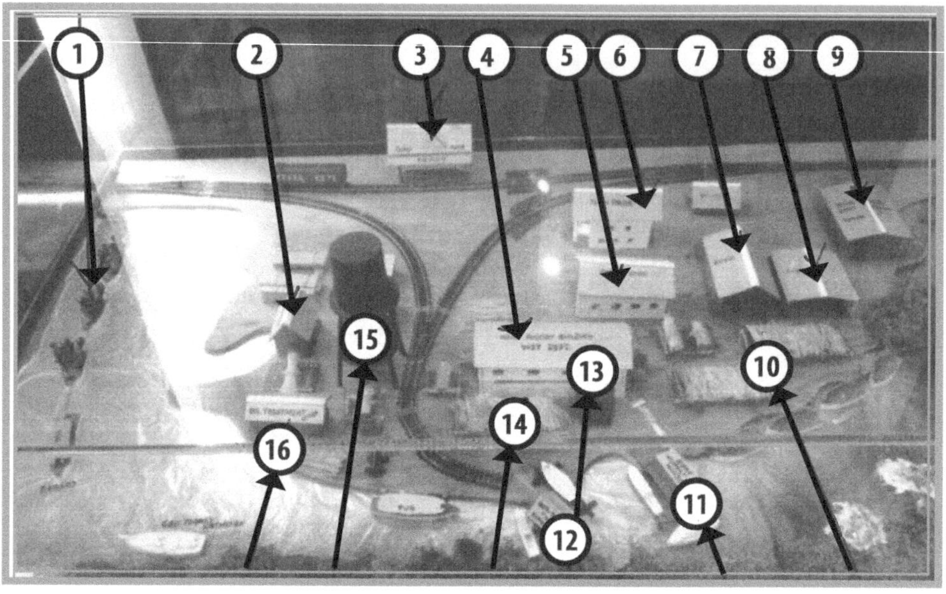


1. Tow boat pulling a load of buds filled barges to the factory.
2. Warehouse
3. Railroad Depot
4. Main Factory Building
5. Andrews House
6. Tooke House
7. Andrews House Garage
8. Metal Shed for Boat Building
9. Broom Room
10. Drying Yard
11. Boat Supply House
12. George Tooke Fish House
13. Boiler Room
14. Wood Pile (fuel for the boiler)
15. Water Tower
16. Oil Treatment House

# HARVESTING THE TREES (BUDS)

Cabbage palms grow abundantly all over Florida and especially along the Gulf Coast. The coastal areas that were used ranged from the Suwanee River to the north down to the Crystal River to the south. Some of the landowners included The Joseph Dixon Crucible and Iron Company, St. George Fechting, William R. Coulter and others. The largest supply came from Patterson-McInnis Lumber Company land in Gulf Hammock, a large area drained by the Waccasassa River.

Only the young trees were used. These were called "buds". After trimming the fronds, they ranged in length from about five feet to eight feet and weighed from seventy five to one hundred twenty five pounds each. The same territory could be cut every three to four years because the palms that were too small were allowed to grow up to a usable size. Therefore, there was never a shortage, in contrast to the Cedar Pencil Industry which depleted its resources within about 35 years due to clear cutting the cedar trees and not replanting.

The Standard Manufacturing Company did not own the land. It leased the right to cut the buds and paid the land owners "stumpage", meaning so much per tree. In 1915-25 this ranged from ½ to 1½ cents per tree.

During the 1930's, 40's, and 50's almost all of Gulf Hammock land located just south of Cedar Key and drained by the Waccasassa River was owned by the Patterson-McInnis Lumber Company. During that period the bulk of the buds came from that area.

*Dr. John cutting a cabbage bud.*

I remember taking trips with my Dad to the town of Gulf Hammock to negotiate leasing rights to cut the buds on their land. As I recall, Highway 19 from Otter Creek to Gulf Hammock was a dirt road in the 1930's. I would sit on the porch of the old Gulf Hammock Hotel and wait for my Dad to finish conducting his business with a Mr. Robinson who represented the lumber company. Our family rarely got to go out of town so I was happy to go with my Dad. We usually saw a lot of wildlife on the way.

*Dr. John holding a freshly cut cabbage bud.*

The company contracted with cutters and paid them so much for each bud. The buds were cut with a special thin bladed axe made by the Collins Axe Company located in Collinsville, Connecticut. Each corner of the blade was sharpened by the cutter so that deep cuts could be made into the trunk of the tree and the blade would cut its way back out after each cut. Otherwise binding would occur making it more difficult to pull the blade out after each cut. A skilled cutter could cut 100 buds per day.

Some of the bud cutters I remember include Sylvester Weeks, Jesse and Pete Ford, Talmadge and Will Watson, Raleigh Hudson, Harry and Ralph Ingram, Fred, Crill and Jewel Kirkland, Charlie and Oliver Miller, Henry Gibson, Charlie Ford and Albert McCain. Only oxen could be used to pull the wagonloads of buds to the river and creek banks because of the wet, muddy terrain. Some located in less swampy areas were trucked out and brought to the factory.

*A mural by artist Bill Roberts depicting the bud cutters camp with Dr. Dan standing beside it. There is also a wagon load of buds being pulled by an ox.* Courtesy: Andrews House Exhibit

Sometimes during the summers when there was an unusual amount of rain or after a tropical storm or hurricane, the swamps became too wet to harvest and transport the buds. The factory would always try to produce more fiber than they could process and sell. The extra dry fiber was stored in warehouses. They could then finish oiling, dry hackling, drafting and cutting this fiber to fill orders for fiber and brushes when they were unable to get the buds from the swamp.

One can only imagine all of the work involved in cutting and getting the buds to the factory. It was extremely hard work and the environment made it more difficult with the heat

*Another mural by artist Bill Roberts depicting cutting and transporting the buds.* In the Andrews House Exhibit

in the summer, with the mosquitoes, ticks, yellow flies and sand flies (no see-ums). Snakes, especially moccasins were there, but I never heard of anyone being bitten. Alligators were also present, but they were not a hazard because they were afraid of humans. It took strong and hardy men to perform this type of work. I doubt that it would be possible today to find those willing to do this work.

Often local kids would bring "boots" from a swamp cabbage to the factory and get paid a dime or so for them. It was probably their father who had cut it for the heart and of course had to remove the boots in order to get to the edible part. The boots were still good for the fiber so the kids were happy to tote them to the factory and get paid for them so they could go to the movie or buy something they wanted.

*Boots from a freshly cut bud.*

An interesting incident arose in the 1930's involving a man who was a convicted felon and was camping in Gulf Hammock on land owned by the Patterson McInnis Lumber Company. He was accused of threatening, and on one occasion, shooting at two employees of the Standard Manufacturing Company. He was also accused of setting fire to the woods causing damage to some timber owned by the lumber company. A letter (see letters at the end of this chapter) was written by Levy County Judge Joe Sales to the Governor of Florida, Dave Schultz, asking that he send a detective to investigate and help resolve the situation.

Another interesting incident occurred in 1917 involving a tidal creek named Demarest located between the Waccasassa and Withlacoochee Rivers. It was being used by the Standard Manufacturing Company to obtain cabbage buds and by the Joseph Dixon Crucible Company to obtain cedar logs. A man who owned the property through which the creek ran claimed to have obtained permission from the U.S. Land Office to put a fence across the creek. Since this was classified as a navigable waterway he was prohibited from doing this (see letter from Collector of Customs in Tampa).

Once there was a stubborn ox that seemed to have a mind of his own and refused to pull a wagon loaded with buds. Despite repeated commands and prodding the worker got so frustrated that he built a fire under the ox. However, the ox responded by moving just far enough ahead so that the wagon straddled the fire causing the wagon to catch on fire.

Another time an ox pulling a wagonload of buds along a creek bank was given the command "gee". Apparently he got confused and instead of turning right he turned left into the creek along with the wagonload of buds.

Another incident involving an ox occurred in town. Occasionally when an ox was not being used in the swamp they were turned loose on Atsena Otie until needed again. My dad got a call from someone on Main Street (Second Street) saying that a black ox named Pete was running loose there. He was retrieved by a factory worker and tied to a tree behind our house. It was my job to feed and water him for a couple of weeks until he was sent back to work in the woods.

*"Pete" the black ox being watered by a young John Andrews near the Andrews House when it was located by the Fiber Factory*

Bronson, Fla., March 3rd 1936

Dear Dr. Dan:-

I am sorry I missed your man Kirkland when he came to Bronson the other day, it was through no fault of mine however, I was close in the office all of the afternoon after your phone call until Mr. Paty came along speaking in his own behalf for Governor, I naturally went to hear him and stayed on the streets chatting with different groups and didnt get back to the office right away after the address, then too I was looking for you personally for that was the impression I had and was not looking for Mr. Kirkland. I also the next day after my Mother told me Mr. Kirkland had been at the house called Mr. McInnis and asked him to have Mr. Ward to look into the situation and both Mr. Ward and Mr. McInnis have just been in the office tonight, just left me at 8.30 PM and I am writing this letter to you right straight. Mr. McInnis and Mr. Ward dont see how they can make the affidavit but will have Mr. Kirkland to come to Gulf Hammock and they will phone me and I will go to Gulf Hammock and have him sign the affidavit there without making another trip to Bronson. I am interested in helping you after all most of all because I have known you longer and am not unmindful of the many, many things you, have so kindly and whole-heartedly done for me, and of course want to help Mr. McInnis and his folks for the same reason. I am sending the Governor of this State a letter, a copy of which is sent you herewith and an additional copy is being sentto Mr. McInnis or handed to him rather, I think the situation justifies the granting of the request which I have made and lets hope we can settle this situation for once and for all.

I want you to know which does not have to be said, that I am at your service and at your command, I will respond and go all the way with you.

With all good wishes, I will let you know what I hear from the Governor.

By the way Dr. Dan you dont have to get others to contact me for what you want from me, the idea of you needing an Aaron in calling me.

Most cordially,

J. C. Sale

*The letter from County Judge Sale to Dr. Dan regarding trouble in Gulf Hammock.*

Bronson, Florida, March 3rd 1936

Hon. Dave Sholtz,
Governor,
Tallahassee, Florida,

Dear Governor Sholtz:-

We have a vexing situation in this County which very vitally is affecting two of the most substantial industries we have, namely the Paterson-McInnis Lumber Company of Gulf Hammock, and the Standard Manufacturing Company of Cedar Key, the first, one of the largest saw mills in the State owning a large percentage of the domain of the County, the latter distinctive in that they are the most substantial and possibly the only successfully operated fibre plants in the State. I refer to a man named J.A. Glaze who lives in the great Gulf Hammock, camping on property of the Paterson-McInnis Lumber Company. This party was at one time convicted of a felony and served a sentence in the State Prison of Florida, was paroled and due to his irritable nature and domineering disposition the clemency was withdrawn and he had to be remanded to the State Prison for the remainder of his sentence. On October 13th 1934 this man was before me as County Judge on peace warrant proceeding, a number of witnesses were heard, to be exact ten of them, sufficient evidence was presented to warrant his being held under bond and he served three months, the maximum provided by law, and was naturally then released, this drew out of threats against employees of the Standard Manufacturing Company who were getting out raw material in the woods; On January 29th of this year two employees of the Standard Manufacturing Co., made complaint before me that on the 25th day of January 1935 J.A. Glaze did shoot at them with a deadly weapon and did attempt to kill and murder them, bond was given for his release in the sum of $1,000.00 signed by substantial men of the County and he naturally was released from custody and at this time is at large. The latest development of this situation is that considerable territory and land of the Paterson-McInnis Lumber Company have been burned, some timber damaged and palmetto buds of the Standard Manufacturing Co., destroyed in the woods between Gulf Hammock town and Cedar Key, and it is alleged Glaze admitted to a witness he did the burning which is a violation of law. The point I wish to make is that I am sure the two business firms and the Sheriff and myself will appreciate you as Governor sending a detective down here to look into this situation and assist in threshing out the truth, and I trust you will do this if at all possible. I am sending or handing a copy of this letter to Mr. A.E. McInnis President of the Paterson-McInnis Lumber Company, and Dr. D.A. Andrews, President and Manager of the Standard Manufacturing Company. I am convinced of the wisdom of asking this and I am sure these Gentlemen will agree with me.

With assurance of my high esteem,

Yours most cordially, [signature] County Judge

*A letter from County Judge Sale to Governor Sholtz regarding trouble in Gulf Hammock.*

UNITED STATES CUSTOMS SERVICE,
OFFICE OF THE DEPUTY COLLECTOR, IN CHARGE.

SUBPORT OF Cedar Keys, Fla.
July, 6th, 1917.

Collector of Customs,

Tampa, Florida.

Sir:-

Subject: Obstruction of Demorest Creek.

Complaint has just reached this office through the Standard Mfg.Co., of this city, that Mr.H.A.Stephens of Inglis, Florida, has advised this company and others that he intends to put a fence across Demorest Creek which will obstruct navigation on this stream.

For your information will say that Demorest Creek lies between the Withlacoochee and the Wacasassee rivers, and is navigable some six or seven miles carrying 3 feet of water. This stream has been in constant use for years by The Standard Mfg.Co., The Joseph Dixon Co., and by many of the net fishermen along the coast. Considerable time and money have been spent by local parties to facilitate navigation on this stream as it affords the means of marketing from a vast section of country such raw material as palmetto fibre, hard wood, cedar logs & etc.
The Standard Mfg.Co. having at present 25,000 palmetto buds at the head of this stream, which will be a total loss amounting to hundreds of dollars, should this channel be obstructed in the manner in which Mr.Stephens proposes to fence it.

Mr.Stephens has recently taken out through the United States Land Office a homestead through which this creek runs about two miles from its mouth. He claims to have been given permission by the Land Office to fence this creek.

*Letter regarding obstruction of the creek.*

# TRANSPORTING TREES (BUDS) TO THE FACTORY

The factory had its own fleet of barges that were towed up the creeks and rivers along the coast to bring the buds to the factory. They were hauled in wagons pulled by oxen and stockpiled on the river and creek banks. Because of the wet, muddy terrain only oxen could be used for this job. It usually took three to four weeks for the cutters to harvest and stockpile enough buds to warrant a trip to get them. The hearts could not be used because this part of the tree had spoiled by the time it reached the factory.

*Bob Stapleton ran the tow boat Maude C. from the 1920's through the 1940's.* Courtesy: Andrews House Exhibit

Mr. Bob Stapleton was the boatman for many years. He knew the coast and tidal creeks like the palm of his hand. A deep-drafted diesel powered towboat (Maude C) about

40 feet long towed the barges to near the mouths of the creeks. The creeks were extremely shallow and rocky so the barges were flat bottomed making them shallow drafted. A small flat-bottomed boat with a tunnel drive called the Tuglet was used to tow the barges in and out of the creeks. The Maude C had a 60 horse diesel engine and the tunnel drive boat had a 28 horse Scripps marine engine. Once out of the creek this tunnel boat and the loaded barges were hooked to the Maude C and towed back to the factory. Only during the height of the spring tides was the water deep enough to navigate the creeks. Typically about five barges were towed up the creek and loaded one day and towed out on the top of the tide the next day.

A crew of two men in addition to Captain Bob were needed to navigate the creeks and load the barges. The creeks were so crooked the two men used long poles to push the barges free when they would hang up on the banks of some of the sharp turns. The barges were tied very close together so that the

*The Maude C. towboat*

men could jump from one barge to another as necessary to free them from the banks on the sharp turns. Some of the creeks from which the buds were retrieved were Halls, King, Jacks, Sheepshead, Double Barrel, Ten Mile, Cow, Demory, Barnett, McCormick, Trout, Dry, and Kelly Creeks.

During the winter months from November through February, the higher of the two daily tides occurred at night instead of during the day. Therefore, the daytime tide was never high enough for the boat and barges to get in and out of the creeks. It is amazing that Captain Bob could navigate the coast and these creeks even on the darkest night with nothing but a compass and a spotlight. He did put down some makeshift markers on the oyster bars around the mouth of the creeks to help him maneuver around the oyster bars.

The barges were secured as close to the banks as possible where the buds were located. This was done with ropes attached from the barges to trees on the bank. There would still be a short distance from the bank to the barge. Wooden gangplanks were used as a walkway that the workers carrying the buds would have to negotiate in order to drop buds into the barges. It required strong men with a good sense of balance to keep from falling while carrying the heavy buds.

Since it was not possible to get all the barges loaded before the tide went out, it was necessary to spend the night and come out of the creek with the tow of barges the next day in the summer, or the next night in the winter. Mosquitoes and sand flies (no see-ums) were terrible. Captain Bob and the two workers slept in the towboats or on the bank. It was tough sleeping in the summertime when it was so hot and with the cabin closed up to keep out the insects. Kerosene was used with a spray gun to try to keep the insects away.

*A barge loaded with cabbage buds.*

Feed in one hundred pound bags for the oxen was delivered by Captain Bob to the cutters on the bud trips. I recall two feeds that were used: cotton seed meal and cotton seed hulls.

Occasionally a contractor would quit after having been advanced money and provided with oxen. In such instances the oxen would be brought back to the factory in one of the barges and tied to trees at our house. It was my job to keep them fed and watered. Sometimes the oxen would be released on Atsena Otie and kept there until needed again. Once a big black ox named Pete swam across the channel to town. He was caught on Second Street, brought back to the factory and returned to Atsena Otie. I think he stayed there after that until he was needed again by one of the bud cutters.

*Tow boats Maude C. and Margaret with tow of loaded barges tied up at the "bud" dock. NOTE: The cabbage palm log pilings supporting the dock.*

*Towboat Margaret before towboat Maude C.*

*This Andrews House Exhibit, a mural by artist Bill Roberts, depicts the towboat Maude C. towing buds to the factory.*

1. Barge with buds
2. George Tooke Fish House
3. Maude C.
4. Boat Supply House
5. Main Factory
6. Water Tower
7. Andrews House
8. Broom House

# THE PROCESS

The mill was steam operated as were most mills back in the early 1900's. The boiler was fueled with wood which consisted of pine and scrub or black jack oak. These trees grew abundantly between Cedar Key and Rosewood and were brought in by rail or truck to the factory. Mr. Joe Jones and Mr. Fred Arline were the firemen I recall back in the 30's and 40's. Later in the 1940's a changeover to oil (Bunker C) for fuel was made. This was stored in a large tank remote from the factory and piped to the boiler in the boiler room.

There was a concrete walled ditch leading from the channel to the back of the factory building into which the barges full of buds were pushed. Here they were unloaded by hand and, with the help of an electric conveyor, were deposited onto the loading dock. This elevated platform was attached to the back of the factory and contained two large cypress cooking vats about ten feet in diameter and twelve feet deep. The first step in the process of extracting the fiber from the buds was to boil the buds.

Handling these buds required very strong men because they weighed from 75 –125 pounds. Laurie Baker was such a man about 6′2″ and 220 pounds, all muscle. He was called the "Tankman". The buds were handed to him and he deposited them into the tanks. The tanks were then filled with water and the buds were boiled for 72 hours. Heat came from steam pipes in the bottom of the tanks. After 72 hours the lids were lifted off the tanks and the buds removed. This was done by Laurie Baker who I remember very well. He would get in the tank and throw them out on the wooden dock that surrounded the tanks. When this was no longer possible he would climb out of the tank. Then with a long wooden pole with a hook on the end, he would snare each bud and, hand over hand hoist it from the tank. A piece of the bottom of one of these tanks can be seen mounted above

*Sample of Cabbage Palm, in the Andrews House Exhibit*

the gate to the Dorothy Tyson Memorial Garden on Second Street. Another piece is over the gate in the fence facing the Andrews House on D Street.

The loading dock containing the cooking tanks was an extension to the north end of the factory building so the boiled buds could be sent through a chute to the lower floor of the factory where the machinery was. The factory was a two story metal building about 125 feet long and 40 feet wide with a main line shaft running its length. Machines operated off belts and pulleys both upstairs and down stairs.

The purpose of boiling the buds was to soften them up making them easier to tear apart into "boots" which contained the fiber. This was done by the "head hackle machine" that was a rotating drum with teeth in it. It had a metal cover with a large mouth about 16 inches in diameter where the butt of the bud was inserted by the "head hackler". This tore away the binding fibers holding the bud together and allowed it to be torn

*A picture of the hackle machine showing the mouth, an Andrews House Exhibit*

*A hackle machine showing the rotating drum with teeth which removed the waste material from the boots, leaving only the fiber. This is an Andrews House Exhibit.*

apart into boots which were removed layer by layer. The boots were actually the bases of the fronds that had been trimmed off of the palm. Then the boots went through a boot roller (similar to a wringer on a washing machine) that crushed and flattened them making it easier to feed them into the "hackle machines" and for these machines to extract the fiber from them. The hackle machine was another machine consisting of a rotating drum with teeth (smaller than the head hackle) and with a smaller mouth. The worker fed these previously prepared boots into the mouth of the hackle machine first one end and then the other. The hackler would then be left with a handful of fiber. These were called "wet" hackle machines because there was a jet of water directed into the machine that washed the fiber at the same time it was being extracted. These removed all of the woody, pithy material from the boots and this waste material was ejected from the back of these machines. This contained some good fibers so "waste shakers" were hired to reclaim this fiber. They wore waterproof aprons and beat the fiber against their aprons and shook it to remove and save the good fibers.

Before going to the hackle machines, the boots were sorted into those containing coarse, medium, and fine fibers. The outer boots contained the coarse, the intermediate ones medium, and the innermost boots the finest fibers.

The waste material that contained the unusable fiber was rolled in wheelbarrows out to the land adjacent to the factory and spread out over a large area and remained there to deteriorate. Frequently, after school some of us would choose up sides and play tackle football on it. It was soft and spongy and difficult to run on but even though we wore no uniforms, falling on it produced no injuries.

A steam whistle was used to announce the workers starting

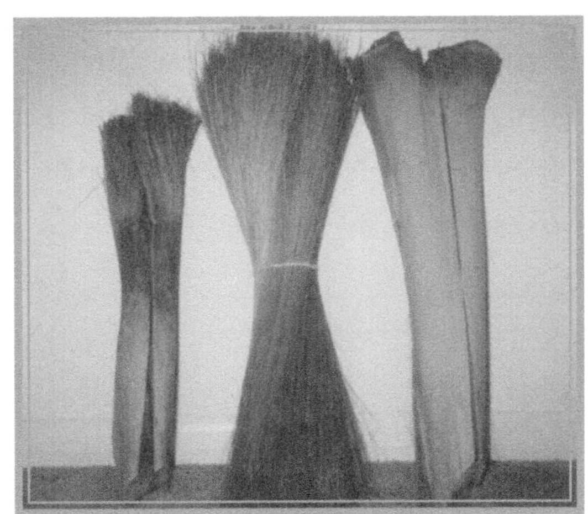

*Bundle fiber with a partially hackled boot on the left and an unhackled boot on the right. There is a bundle of fiber in the middle. This is in the Andrews House Exhibit.*

and quitting times. There were two short blasts at 7:55 a.m. and a single long blast at 8:00 a.m. This was repeated at 11:55 a.m. and at noon and again at 12:55 p.m. and 1:00 p.m. denoting lunch hour. The end of the workday was announced at 4:55 p.m. and 5:00 p.m.

During the lunch hour some of the workers would bring fishing poles and fish for redfish, trout and yellowtails in the channel off the "bud dock". They often caught enough for supper.

The factory had a lumber shed containing wood for repairs and for building barges and skiff boats that were used for getting the buds to the factory. They also made their own wooden wheelbarrows (one can be seen on the front cover of the book). There was a large shed with a metal roof behind our house where the boats were built on a cradle. When finished and the tide was high, using rollers they were pushed to the water's edge and launched. The lumber used was cypress and came from either the Tilgham Cypress Company in Lukens or the Patterson McInnis Lumber Company in Gulf Hammock. The cypress wood swelled a great deal when wet closing the seams in the bottom so no caulking was necessary.

## DRYING AND OILING THE FIBER

The next step in the process was to dry and oil the fiber. Small bundles of the wet fiber was crisscrossed on upright wooden stands and wheeled with dollies out to the drying yard adjacent to the factory. The yard consisted of long lines of elevated chicken wire frames on which the fiber was spread out to dry. There were cement walkways between the rows of frames making it easy to operate wheelbarrows and dollies. There was a hinged lid that kept the wind from blowing the fiber away. On a sunny day it took about 3 hours for the fiber to dry. Sometimes when the fiber was dry and a rain shower "squall" was coming, many workers

*Drying yard in the foreground where fiber was sun-dried. Waste fiber is in the background.*

would leave their posts and run out to take the fiber up before it could get wet. Sometimes even non-workers that were around and saw what was going on would stop and help.

Once it was dry it was tied into bundles about 5 inches in diameter. It was loaded into wheelbarrows and wheeled to the oil treatment house.

The oil treatment house was a one-story metal building about 30 feet wide and 80 feet long. The purpose of the oil was to help preserve and keep the fiber from being brittle.

Three workers were required to perform this operation. One loaded the dry bundles into small metal racks with wheels. Another lowered these racks with a hoist into a vat of paraffin oil (a petroleum based oil purchased from the Gulf Refining Company named Cayuga Oil). The third worker rolled the carts of oiled fiber up the ramp that was attached to the reservoir. It was left there for 4-6 weeks so the excess oil was allowed to dissipate so it would not be too oily when it reached the finishing department.

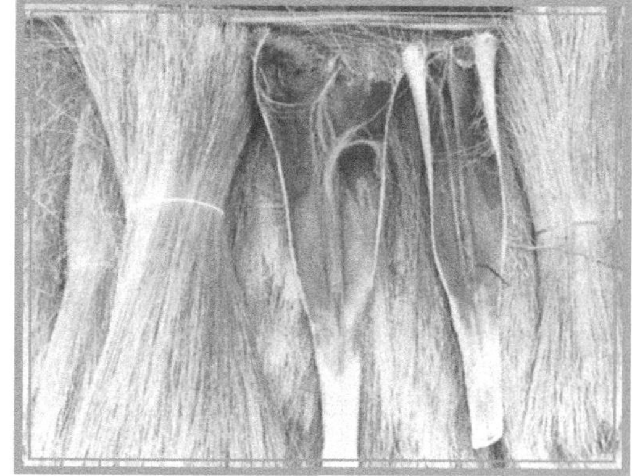

*Bundles of dry fiber and a couple of boots. This is in the Andrews House Exhibit.*

# THE FINISHING FIBER DEPARTMENT

The finishing department was located in the main factory building. There was a long wooden ramp accessing the south end of the factory building allowing wheelbarrow loads of the oiled fiber from the oil house to reach the finishing department that was located on the second floor of the main factory building.

During the drying and oiling process some of the fibers in the bundles had become crisscrossed and tangled. The first step in the finishing department was to use a "dry hackle" to straighten it so that all the fibers were then longitudinally arranged.

From the dry hackle it went to the drafting tables where it was impaled between 2 sets of "combs" (a series of 3 rows of slender sharp metal spikes about 5 inches long on wooden boards about 2 feet long and 2½ inches wide) one attached to the table and the other used to impale impaled fiber between the two combs. The drafters would then pull out all the fibers that were the same length from one end and place them in u-shaped wooden holders. Then they would reverse the impaled fiber and do the same thing on the other end. This left all of the remaining fibers almost the same length. This avoided trimming off any more than necessary to be sure all of the fibers in the bundles were the same length. This process was called drafting.

*Matilda Hathcox drafting fiber (separating it into different lengths).*

*An Andrews House exhibit showing fiber impaled in combs ready to be drafted.*

The next step was to fill the many orders from brush-making companies. The factory received orders from companies all over the US as well as some from Canada, England, Germany and even Australia. It was the most suitable fiber available for making all kinds of quality brushes. Orders were received for a variety of lengths varying from 3 ¼ inches to "natural length" which was approximately 20 inches. The different grades of fiber were coarse, medium and fine. Hand cutters designed by Mr. St. Clair Whitman and made by the Maddox Foundry in Archer did a remarkable job of cutting this very tough fiber. My Dad and Uncle Forrest did most of the cutting to fill each order.

The product was called palmetto fiber. It was packed in wooden crates holding 50 to 500 pounds, depending upon

*Belle Sanders cutting and preparing fiber to be sold for manufacture of brushes.*

the order, and shipped out on the train. It was later trucked out after train service was discontinued in 1932 by the University City Transfer Company in Gainesville.

There were certain times that conditions in the swamps along the Gulf Coast made it impossible to harvest the trees for a period of time - such things as hurricanes, tropical storms, or extremely wet summers. The company had three warehouses that were used primarily for storing large quantities of incompletely processed fiber (dry un-oiled fiber) so they could keep the factory operating and fill their customers' orders. The demand for fiber was so great that the factory never closed during the great depression of the 1930's. People living in Cedar Key were also more fortunate than most communities during that time because of the abundance of seafood they could eat or trade for farm products raised in the farming areas of Levy County.

During its years of operation the Fiber Factory was an asset to the town aside from providing employment for 100 people. It served several needs of the town that were not otherwise available. The set of ways was made available to local boat owners who used it to do needed maintenance work on their boats. It maintained a good supply of lumber for sale that supplied local needs and also kept a supply of hardware some of which was not available at the local hardware store. "Mack" McCain always said that if you needed something and had trouble finding it, "go to the Fiber Factory. You will probably find it there".

*A hand cutter and fiber which has been cut to fill and order from a brush company. This is in the Andrews House Exhibit.*

*A hand cutter and cart containing bundles of fiber.
This is in the Andrews House Exhibit.*

# THE DONAX BRUSH DEPARTMENT

Dr. Dan designed a variety of brushes for which he chose the trade name DONAX. Donax is the generic name for coquina which is a small mollusk. The shell is multi-colored and quite pretty. They are found in great numbers along the beaches on the east coast of Florida. Solidified layers of these shells were found on the ocean floor near St. Augustine. This was mined, cut into blocks and used to build many of the old buildings in St. Augustine. The old fort there is also constructed of coquina.

The most popular of the brushes was called a "whisk broom" and Dr. Dan obtained a patent on his unique design of this brush in 1912. It was used for brushing dust and lint from clothes as well as anything else that harbored dust. Men commonly carried a whiskbroom in their back pocket. Another brush that was much smaller was called a "hat brush" and was used for brushing the dust off of men's felt hats. There were two sizes of each of the above brushes.

Other brushes designed by Dr. Dan and made at the factory in Cedar Key included a "crumb brush" used with a dustpan; an "auto brush" used to sweep out the floorboard of cars; as well as some hearth brooms. A metal handled utility brush was the cheapest brush made since it was not handmade like the rest of the above brushes. The handmade brushes consisted entirely of fiber, even the handles.

*A cutter used to trim crumb brushes. This is in the Andrews House Exhibit.*

I will try to describe how the whiskbroom was made. The fiber was first cut into two tufts, each of a different length, the longer one to form the handle. This longer part was

bound to form the looped handle. Below the handle this tuft of fiber was divided to encompass the shorter u-shaped tuft of fiber. A heavy red cotton thread was used to bind the fiber in proper position. The handle was then dipped in shellac, requiring four coats. A day of drying was required between each coat and they could not be dipped on days when the humidity was greater than 80% as the shellac would become opaque when dry instead of clear. This would decrease their value. After the shellac had completely set, the handles were as hard as wood.

*Set-up for drying brushes after they have been dipped in shellac. This is an Andrews House Exhibit.*

These brushes were made in the building called the "BROOM ROOM". My aunt Agnes Tooke (my mother's sister) was superintendent of the Brush Department where she and six or seven other ladies worked, making the brushes. They included Mrs. Mattie Andrews (no relation), Mrs. Laura Byrd, Mrs. Madeline Richard, Mrs. Hattie, Mrs.

Nellie Whitman, Mrs. Lizzie Depew, and Mrs. Ida Depew. Mrs. Laura Jean Delaino worked there one summer while out of school combing the fiber that straightened it out preparing it for the ladies to turn it into brushes.

They worked at a long table in an assembly line fashion since there were several steps involved in making each brush. One of these tables is now in the museum in the

*An Andrews House Exhibit showing a display of brushes.*

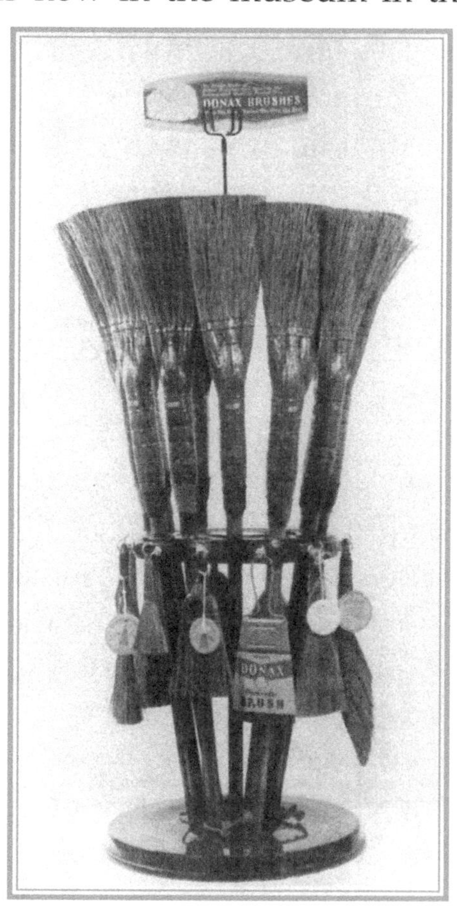


*Another Andrews House Exhibit showing a display of brushes.*

Andrews House Factory Exhibit where it is used for some of the displays.

The Brush Department created a beautiful exhibit of these artistic brushes and entered it in several World Fairs including Cleveland, Chicago, and New York. The exhibit consisted of artificial palm trees with brushes hanging from them. There were beautiful wooden stands with all of the different kinds of brushes hanging from them along with a number of photographs. The Standard Manufacturing Company received a Certificate of Merit for this Florida Exhibit in the 1939 New York World's Fair. It is now displayed on the wall in the Fiber Factory exhibit in the Andrews House. Unfortunately the display itself was largely destroyed in the 1950 hurricane named "EASY" which destroyed the Broom Room building.

The brushes were widely sold over the United States and throughout Florida. The Marshall Field's Department Store in Chicago was also a major customer. Marshall Field's Department Store in Chicago was a major customer. They were sold in many stores and gift shops all over Florida. A number of larger cities in Florida bought them, fitted them with leather cases with the name of the city embossed on them and used them to advertise and promote their cities.

A copy, almost identical to Dr. Dan's patented whiskbroom, was made and sold in 1917 by the Palmetto Whisk and Broom Co. in Jacksonville, Florida. Dr. Dan filed a complaint with the patent office in Washington D.C., as well as with the Unfair Trade Act office in Jacksonville. However, after obtaining legal advice he decided not to pursue this because of the potential cost of the litigation.

There was some fiber left in one of the warehouses after the factory closed in 1952 which my mother put in her garage and saved. I now make replicas of some of the brushes. I donate them to the Cedar Key Historical Society Museum where they are sold to help support the museum. I worked in the factory as a teenager drying and oiling the fiber and

*A tray made by Agnes Tooke (my aunt) and more brushes from the factory.*

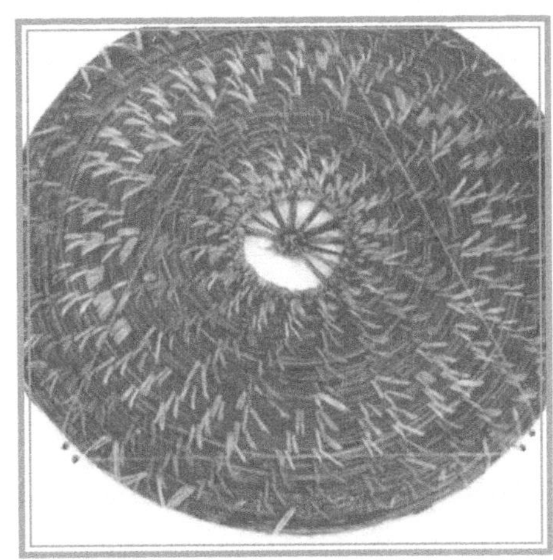

*Using medium fiber with raffia, Agnes Tooke made four of these mats as a wedding gift for Etta (Wadley) Webb Watson in 1938. This is an Andrews House Exhibit.*

also went on several of the boat trips to get the buds and bring them back to the factory. However, I never made or even watched the ladies making the brushes so I had to figure out how they were made. After considerable trial and error I learned out how to make the loop for the handle of the Donax whisk, and I now make those as well as the hat brushes. I have also taught a niece and nephew how to make brushes so the art will not be lost until the fiber runs out. I have quite a supply of it but after it is used up there will be no more. The cabbage palm still grows abundantly in Florida but to extract the fiber is such a complicated, expensive, labor-intensive process that it is unlikely it will ever be produced again.

## MARKETING THE FIBER AND BRUSHES

In the early years of operation of the factory, Dr. Dan made numerous trips around the country showing samples of his fiber to prospective brush company customers. In a very short time he developed a large market for the fiber as well as for the Donax brushes. These included companies in cities all over the United States and also companies in Canada, England, Germany and even one in Australia. Ninety five percent of the fiber produced was sold to these companies that used it to make brushes of various types. Only about 5 percent of the fiber produced was used to make the Donax brushes. The fiber itself was therefore the major product of the Standard Manufacturing Company.

In 1918 the fiber sold for 20 to 25 cents per pound depending upon the grade. The finest of the three fiber grades was less durable than the medium and coarse grade and therefore was the cheapest.

There was never a shortage of orders for the fiber which allowed the factory to operate continuously during the great depression of the 30's. This was the highest quality fiber available for the manufacture of all types of brushes except sweeping brooms. Those were made from broom corn fibers which were much cheaper than palmetto fiber. Palmetto fiber in terms of durability, color and varying degrees of stiffness was the most desirable fiber available and was produced only in the United States and specifically in the three plants in Florida. One of the other two was in Jacksonville and the other in Sanford.

There were 2 imported fibers that were used to make brushes, palmyra fiber from the palmyra palm in India and tampico from the agave plant in Mexico. Both of these were inferior in quality to palmetto fiber. The Fuller Brush Company made a wide variety of brushes. Some were sold door-to-door by the "Fuller Brush Man". However, I have not found a record of them using The Standard Manufacturing Company's palmetto fiber. They are still in business and produce and sell brushes made from synthetic and possibly some from palmyra and tampico fibers.

# FIBER AND BRUSH COMPANY CUSTOMERS OF THE STANDARD MANUFACTURING COMPANY

**(July 1931 – June 1932)**

## FIBER

- Atlanta, Brush Co., Atl. GA
- Standard Brush and Broom, Portland, IN
- J.I. Holcomb Mfg. Co., Indianapolis, IN
- American Brush Co., Chicago, IL
- Detroit Quality Mfg. Co., Detroit, MI
- U.S. Brush Co., Omaha, NE
- Aurora Brush Co., Aurora, IL
- S.A. Felton & Son Co., Manchester, NH
- Opie Brush Co., Kansas City, MO
- Manufacturers Brush Co., Cleveland, OH
- Mann Brush Co., Montreal, Quebec
- Michigan Brush Co., Detroit, MI
- Flour City Brush Co., Minneapolis, MN
- Joseph Flatt Co., Reading, PA
- Northwest Manufacturing Co., Omaha, NE
- Joseph Minck Co., New York City, NY
- Ernzer Manufacturing Co., Burbank, CA

- Sullivan Brush Co., Terre Haute, IN
- Charles Feldstein Co., Philadelphia, PA
- Standard Brush Co., St. Louis, MO
- Leninburg Brush Works, Chicago, IL
- Eaten, Schlick & Wall Co., New York City, NY
- Milwaukee Brush Co., Milwaukee, WI
- Palmetto Products Inc., Waukesha, WI
- W.W. Leonardi Co., St. Augustine, FL
- Young and Schwartz Co., Buffalo, NY
- Warren Brush Co., Cincinnati, OH
- Union Brush Co., Cincinnati, OH
- Maurice Beitz Co., Chicago, IL
- Western Brush Co., Chicago, IL
- American Brush Mfg. Co., Portland, OR
- American Express Co., Toledo, OH
- Paul Bennet Co., Royal Oak, MI
- Creamery Pkg. Mfg. Co., Chicago, IL
- Hoase Brush Co., Claremont, CA.
- California Brush Co., Los Angeles, CA
- Maryland Fiber Products Co., Baltimore, MD
- Schaefer Brush Co., Milwaukee, WI
- Sing Sing Prison, Ossining, NY
- A.W. Justman Brush Co., Omaha, NE

- Workshop for the Blind, Salt Lake City, UT
- Richards Brush Co., Seattle, WA
- Osborn Mfg. Co., Cleveland, OH
- B.C. Brush Works, Vancouver, B.C.
- Empire Brush Works, Chicago, IL
- Florida Brush Co., Tampa, FL
- A. Laitner and Sons, Detroit, MI

## BRUSHES

- Fred Anderson Co., Kilbourne, WI
- George A. Klein, Co., Muncie, IN
- Marshall Field & Co., Chicago, IL
- Dean Anderson Co., Ashville, NC
- L.S. Donaldson Co., Minneapolis, MN
- The Killian Co., Cedar Rapids, IA
- H.L. Hymes, New York City, NY
- H. R. King Co., St. Petersburg, FL
- B. Shapiro Co., Jacksonville, FL
- Herzfield Phillipson Co., Milwaukee, WI
- Reed-Cook Co., Camden, NJ
- Reich-Ash Co., New York City, NY
- Kirby, Block & Fischer Co., New York City, NY

- Conner & Conner Co., Wabash, IN
- John Barlow Co., Lawrence, MA
- Asher, Cushing & Foster Co., Boston, MA
- Merchandise Buyers, Inc., New York City, NY
- Venice Curio Store, Tarpon Springs, FL
- Rinnette Gift Shop, Lakeland, FL
- Chris Gift Shop, Lakeland, FL
- G. A. Hutchinson, Co., Hollywood, FL
- S. R. McIntosh Co., St. Petersburg, FL
- Vanity Fair Gift Shop, Lakeland, FL
- Steves Curio Shop, St. Petersburg, FL
- Tampa Alligator Farm, Tampa, FL
- L.S. Hutchinson Co., Philadelphia, PA
- Brown's Shop, Sarasota, FL
- Lillian Shop, Miami, FL
- Standard Brush & Broom Co., Portland, IN
- Betty Wilson Gift Shop, Miami, FL
- Opportunity Gift Shop, Miami, FL
- Rox Stationery Co., Pensacola, FL
- The Treasury Chest, Daytona Beach, FL
- Japanese Gift Shop, Miami, FL
- Foster & Reynolds, St. Petersburg, FL
- National Candy Co., St. Louis, MO

- Paul Bennett Inc., Royal Oak, MI
- The Fair, Chicago, IL
- Barrett Heide Co., Joliet, IL
- H. A. Triplett Co., St. Joseph, MO
- Frankfurth Heide Co., St. Joseph, MO
- L. A. Rosenthal Gift Shop, Charleston SC
- J. B. Ivey & Cc., Charlotte, NC
- E. A. Bugwell Co., Milwaukee, WI
- Eddy's Studio, Pines, NC
- Eckerds of Raleigh, Raleigh, NC
- Barnhill Art & Camera Co., Boulder City, NV
- Miller & Rhoads, Inc., Richmond, VA
- Ames & Brownlee, Norfolk, VA
- Scranton Drygoods Store, Scranton, PA
- Goldblatt Bros. Department Store, Chicago, IL
- Wiebelt Stores, Inc., Evanston, IL
- O. K. Sales Co., Ashland, KY
- Victor Brown Co., Minneapolis, MN
- John Taylor Drygoods Co., Kansas City, MO
- J.C. Spare Co., Schuylkill, PA
- August Bull Candy Co., San Antonio, TX
- J. L. Brandeis & Sons, Omaha, NE

- Halbash-Schroeder Co., Quincy, IL
- W. P. Wallace Co., Osweego, NY
- Baum's Variety Shop, Alexander Bay, NY
- Florida Seashell Co., Ft. Myers, FL
- Scudder's Antique & Gift Shop, Silver Springs, FL
- Oris Curio Shop, St. Petersburg, FL
- Oriental Arts Co., St. Petersburg, FL
- Neptunes Curio Shop, Tarpon Springs, FL

*A collection of Donax Hat Brushes made by Standard Manufacturing Company in Cedar Key, Florida. Note the leather cases indicating different places they were obtained from: Chattanooga, Tennessee; Lake George, New York; Cedar Key, Florida; Empire State International Exposition, 1939 (probably used as souvenirs).*

Photo Courtesy: Bill Bale Collection

# OFFICE OF THE STANDARD MANUFACTURING COMPANY

The office where all the business of the company was conducted was located in the Andrews House. This was our family home and where my brother Dan and I were born and raised. It was in the front room that was part of the original three room house before the additions were made. The rest of the house was our home.

*The office telephone,*
*in the Andrews House Exhibit*

Some of the secretaries and bookkeepers that

I remember were Mrs. Alba Kirchhaine, Mattie Merle Hodges, and Eudora Boothby. My aunt Virginia Tooke (my mother's sister) came to work there right out of high school in 1925. She was replaced by Eudora Boothby in 1932 after she married and had children.

After her husband, Dr. James Turner died in 1940, Aunt Virginia returned and worked there until the factory closed in 1952. She became an excellent typist and developed her own version of shorthand which served her very well for taking dictations from Dr. Dan.

*Mrs. Virginia's work area including original desk, chair, typewriter, adding machine, and check protector. This is in the Andrews House exhibit.*

Her many duties included handling all of the company's correspondence (and it was voluminous) as well as keeping the company's books, making weekly bank deposits and preparing the payroll each week for the nearly 100 workers. One interesting thing I discovered in going through some of the old ledgers was that during World War II, Congress passed a "War Tax" of 5 percent on wages that exceeded $12 per week. This was to help pay for the cost of the war. She also kept the Andrews Brothers books (this was a separate company that owned some Cedar Key real estate). Dr. Dan was also one of the owners of the first power plant in Cedar Key and she kept those books as well. Western Union was also located in the company office during World War II and she and my mother were required to deliver messages to soldier's families.

She also kept the books for the Island City Investment Company. This was formed in 1946 to sell refunding bonds to raise money to settle with the Bonding Company which was threatening to foreclose on the city of Cedar Key for defaulting on a $150,000 bond issue the city floated in 1927 to pave the streets and install the water and sewer system. This was successfully settled with the Ben Hur Life Insurance and other bondholders in 1952.

At the same time my Aunt Virginia was holding down this highly responsible and hectic job, she was a single mother raising and educating her two daughters, one of whom became a school teacher and the other a registrar at a college in Jacksonville. Looking back it is hard to imagine how she could have performed all of those tasks even if she had had a college education.

## EMPLOYEES

The Fiber and Brush Factory was a very important contributor to the economy of Cedar Key during the years of its operation as it employed approximately 100 workers.

The following is a list of names and jobs of some of these workers that I recall or found in the old ledgers.

## BUD CUTTERS

- Charlie, Jesse and Pete Ford
- Will Watson
- Ralph and Harry Ingram
- Fred, Dell, Jewell, and Crill Kirkland
- Will Richardson
- Sylvester Weeks
- Albert McCain, Father of Mac McCain and Father-in- Law of Thelma McCain
- Steve Gore
- Henry Gibson

- Charlie Miller, Father of Oliver
- Oliver Miller, Father of Barbara Miller McJordan

## WET HACKLERS

- Richmond Stewart
- "Punch" Stewart
- Daniel Stewart
- Charlie Mitchell
- David Mitchell
- Charlie Blanche
- Joe Blanche
- Bob Demps
- Abron Demps
- Frank Jenkins
- Charlie Pinkney
- John Aldridge
- BOATMEN
- John Tooke – my mother's brother
- George Tooke – my mother's brother
- Bob Stapleton – Marion Wilson's uncle
- Arthur Campbell

## FIREMEN FOR THE BOILER

- Joe Jones
- Fred Arline (Father-in-Law of Grady McLeod)
- Lonnie McElroy
- DRYING YARD WORKERS
- Julia Smith
- Cleve Wilder
- Eli D. Fletcher
- B. Fletcher

## BRUSH DEPARTMENT

- Agnes Tooke, Superintendent (my mother's sister)
- Hattie Whitman
- Nellie Whitman
- Bertie Whitman
- Mattie Andrews – (no relation, Jackie Andrews Padgett's Mother)
- Laura Taylor Bird
- Madeline Taylor Richard
- Rosa Stapleton (Marion Wilson's Aunt)
- Rosa McCain
- Lizzie Bishop Depew
- Ida Depew

- Helen Kirkland
- Laura Jean Delaino

## FINISHING DEPARTMENT

(Where orders were prepared for shipping)

- Belle Sanders
- Matilda Hathcox
- Rosa Osteen
- Forrest Andrews
- Dan Andrews, Jr.
- Dr. Dan

## CARPENTERS

- Roy Walker
- Mitch Wilder
- Nardie Wilson

## MISCELLANEOUS JOBS, OR WHERE NEEDED

- George Foxwell
- Marion Sexton
- Cleve Wilder

- Leslie Richburg
- Mellious Richburg
- Sammy Lindsey (Lindon Lindsey's Brother)
- Wilson Walker
- Collie Ingram
- Laurie Baker (Tank Man - loaded and unloaded buds from boiling tanks)
- Johnny Andrews (myself)

## SUPERINTENDENT OF WET DEPARTMENT –

- Mr. St. Claire Whitman (Where hackling of boots to produce fiber took place)

He was also in charge of maintenance of all machinery. This included the steam engine, the head hackle machine, the boot roller, the wet hackles for extracting the fiber and all of the hand cutters that he designed.

## BOOKEEPER AND SECRETARIES

- Alba Kirchhaine
- Virginia Tooke Turner Pugh (Sally Baylor's mother)
- William Tooke (my mother's brother)
- Tim Murphy
- Eudora Boothby
- Mattie Merle Hodges

STANDARD MANUFACTURING CO., INC.

MANUFACTURERS OF

ALL GRADES OF PALMETTO FIBRES

LONG STOCK OR CUT TO EXACT LENGTH

CEDAR KEY, FLORIDA

*Fiber Factory stationary*

## EXPENSES

**Buds** ................................ 7 to 15 cents each in 1919

**Paraffin Oil** ............................... 15 cents per gallon (210 gallons per month)

**Stumpage** ............... (to landowner) ¾ cent per bud

**Cutters** (1917)....................2 ½ to 5 ½ cents per bud

**Gas**............................................ 23 cents per gallon

**Shellac** .......................................53 cents per pound (This was bought in 125-pound containers)

**Labor** ................................ 7 ½ to 15 cents per hour

**Wood for boiler** ............................... $2.50 per cord

**Office workers** .............................30 cents per hour

**Unskilled labor** ......................... 7 ½ cents per hour

**Skilled labor** ...............................15 cents per hour

**The wet hacklers**.........................4 cents per pound

**Finishing Department** ................15 cents per hour

**Feed for Oxen**...... Cotton seed hulls $ 1.50 per bag

**Cottonseed meal** ............................... $ 3.25 per bag

*Front of promotional company literature*

*Back of promotional company literature*

WILL LAST A LIFETIME

# PALMETTO FIBRE

is the acme of durability in the Brush World, with a market value of many times that of broom corn.

## DONAX-WHISKS

are made exclusively of Special Selected Hand-Picked Palmetto Fibre.

¶ In offering our DONAX-WHISK we have every reason to believe we are introducing the best, most durable, and artistic Whisk Broom ever put on the market.

¶ The handle will not become loose, nor will the fibre stub off like the brooms you are used to using.

*Inside of promotional company literature*

WILL LAST A LIFETIME

## DONAX-WHISKS

will not wear out in a short time, but will stand the hardest kind of usage and still be just as good as ever.

¶ When you have purchased a DONAX-WHISK you have the broom game beat. Cost a little more, but they are worth it.

When the dust is on the garment
And the lint is on the sleeve,
The active use of this Donax broom
Will the evil soon relieve.

The Palmetto is the fibre
Best suited for this "stunt,"
As it will last a lifetime
And always bear the brunt.

# HISTORY OF THE ANDREWS HOUSE

When Dr. Dan purchased the Fenimore Mill Site from Mr. Hodges, there were a number of houses (approximately 15) that had survived the 1896 hurricane and had been moved from Atsena Otie to this property. Mr. Hodges moved those houses except two to other properties he owned so the Fiber Factory could be built. One was called the "long house" to which a second story was added and it became the "broom room" where all of the brushes were made. The other was a small

*The beginnings of the Andrews House*

three-room house that Dr. Dan retained for the Fiber Factory office and his residence.

After Dr. Dan married my mother he made two large additions to it so it was then a large two story house. The front room remained the office for the Factory. The kitchen and dining room were downstairs and our bedrooms and living room were upstairs.

*Andrews House prepared for the final addition.*

*The completed Andrews House*

Also Dr. Dan brought his dental equipment down from Indianapolis and set it up in a small room upstairs. Every Sunday afternoon he did dental work on his family which included the large Tooke family and some of his friends. He filled some of my teeth with the tools that are on display in the Andrews House. If anyone in town had a toothache he would extract their tooth for them. He never charged for his services.

Dr. Dan became a very prominent citizen of Cedar Key. He was much involved in business and civic affairs as well as local politics. He served as President of the Chamber of Commerce, was on the City Council including serving as Mayor, helped establish the first power plant and his

office housed Western Union during World War II. He also served on the Levy County Selective Service Board, the County Commission, and was President of the Cedar Key State Bank.

*Dr. Dan's foot operated dental drill, spittoon and a barber chair which he used for his dental chair. This is in the Andrews House exhibit.*

# GLOSSARY

**BOOT** The base of the palm frond from which the fiber came.

**BOOT ROLLER** Machine used to flatten and crush boots before they go to wet hackle machines.

**BUD** Harvested cabbage (Sabal) palm after the fronds are removed.

**BUD CUTTERS** Workers who cut the buds with an ax.

**BUD DOCK** Dock used for tying up boats and barges and where the boat supply house was located.

**COMBS** Used to impale the fiber for drafting.

**COON OYSTERS** Small oysters (too small to harvest) comprising the oysters bars. So called because raccoons frequent them during low tides looking for small shrimp and crabs that live there.

**DONAX** Generic name for cochina shells selected as the trade name for the brushes.

**DRAFTING** Process of separating fiber into different lengths.

**DRY HACKLE** Machine used to straighten fiber that had become crisscrossed or tangled in the drying and oiling process.

**FRONDS** Locally known as "fans".

**HEAD HACKLE** Machine used to tear the bud apart into boots.

**HEART OF PALM** The edible center of the cabbage palm.

**NATURAL LENGTH** Total length of uncut fiber. Some brush companies ordered this rather than fiber cut to specific length.

**OYSTER BARS** A solid growth of contiguous small oysters forming shallow bars near the Gulf shoreline. They become exposed during low tides and are submerged during high tides. They are numerous and present navigational hazards.

**PALMETTO FIBER** Name used for the fiber.

**PARAFFFIN OIL** A petroleum-based oil used to oil the fiber.

**SABAL PALM (CABBAGE PALM)** Source of the fiber.

**SAND FLIES** or **GNATS** No-see-ums

**SET OF WAYS** Used to pull boats and barges out of the water for maintenance.

**STUMPAGE** Price paid to land owners for the buds.

**TUFTS** Term used for the two small bundles of fiber used to construct the Donax whiskbroom.

**WASTE FIBER** Usable fiber among the waste material expelled from hackle machines.

**WASTE SHAKER** Women workers who reclaimed good fibers from the waste material.

**WET DEPARTMENT** First floor of the factory building where the boots were hackled. The floor was constantly wet because of water used in the hackle machines. This process cleaned the fiber as it was being extracted from the boots.

**WET HACKLE** Machine used to extract fiber from boots.

Made in the USA
Middletown, DE
11 July 2026

29859729R00066